CLOTHESLINE SAGA

poems

NOAH BURTON

Clothesline Saga ©**2020** by **Noah Burton**. Published in the United States by Vegetarian Alcoholic Press. Not one part of this work may be reproduced without expressed written consent from the author. For more information, please contact vegalpress@gmail.com

Cover Art: Patrick Yagow

Author Portrait: Sadie Brent

Table of Contents

Arboretum...11

Canal...12

City Planning...13

Compact Living...14

Lionel...15

The Enjoyments...16

Mortar...17

The Purpose of the Thimble Inside the Cabinet...18

A Foot and a Half...19

The Cartographer Knows The Cryptographer Hid the Island...20

All is Bright...21

Foul...22

Anniversary...23

Fearing Another Break...24

Proper Response...25

What is Passing What Has to Breathe...26

Away From/It Floats...27

Vice...28

Wilted or Not It is the Same Herb...29

"Two Minds."/ "But You Must be One."...30

What is It You Hear?...31

I Always Think I Hold what I Love...32

In Sight of the Furnace..33

"My kin"...34

Though I Long to Be Noah...36

This Kills Me...37
Chlorophyll...38
Holiday Lawmaking...39
Many and Beautiful Things...40
Strength...41
Head in the Clouds...42
The Hardened Garden...43
The Morals of the Cartoon...44
Average Atlantis Map and Description...45
The Hive...46
Dioramas of Bunkers...47
Grounded for No Reason...50
Poachers, He says...51
Love the Woman in the Snow...52
MDF...53
Self Preservation is a Box Duct
Wrapped with a Globe Inside It...54
Sweet Cyprus...55
Range...56
Love Poem...57
Cut...58
It Figures...59
Ache...60
Nightcap...61
Alone and Happy...62
Rocking...63
You Art That and What is Given...64
Cleanliness...65
DIY...67
Personal Statement...68

But if all of our losses are a mirror
in which we see ourselves advance,
I believe in its terrible, empty reflection,
Its progress from judgment toward compassion.

-Jon Anderson, "A Commitment"

"Second Soldier: Wait a minute!
Suppose two swallows carried it together?

First Soldier: No, they'd have to have it on a line."

-Monty Python and the Holy Grail

CLOTHESLINE SAGA

for Sluggo

Arboretum

An echo. A gleaming.
These are the sleepers
passing in winter sun
in their frozen cars.

Eternity's eagle is growing.
Just ahead at the intersection
a woman runs in place
her dog still as a drum
on an island wiped clean
by swords and illness.

For one moment you stared
at a stone. Then another.
Over each, your boots.

Sometimes your life closes
its fist in the light.
It is enough to make a ball
of flesh beat with blood.

The wing's shadow on the crowd's
head is eternity's plucked umbrella.

Canal

What I needed most when the house
became empty, except for my plates
and box of mixed greens, cans
of sardines, a Turkish coffee pot,
was a quiver of arrows—I needed
to strike the oak to hear the thunk
of the same matter on each other
as dogs bark into the thicket
at coyotes howling on the train tracks.
In loss there is a bowl. In gravity
there is always a floor to curl on.
In my brain, a golden heart encased
in incense dustings. To be open
the door must leave its frame.
A long hallway. Skylights. Crawl space.
Walking it, my head walks backwards
into the morning with my hand on your
waist bare and chilled above the wool
along our thighs like unwound equators.

City Planning

Denied a detour through the marsh
after the first quarter of my life,
I picked up an ice pick leaned up against
a pine tree. The past is frozen, I said
to myself—Ok, and I'm hot and ready to swing
this steel into crystallized needles, make a platter
of snow of them. When the cell is full,
the wall is impenetrable to the human
standing outside of it and the human
standing inside. Today I fear this parasite.
I fear the tiny coiling of the skyscraper
over a meager village where false coins are made.
The inward ventilation shafts cave in
over the Windows 95 computers in the back offices.
There is no weather in the heart, no ebbing bath
outside a bamboo hut. What the skin
feels most in the rib cage is a knotting tendon
and under the light the narrow crater where miners
excavate the replica of a more ancient ziggurat.

Compact Living

You dropped your face to my palm and sat
on the curb—two developmental blue prints.
Then you heard the cement truck roll by
like a wet pile of leaves in a giant's bread basket:

The shovel? He uses it as a spoon.
Around the corner a lamb leaned
against a realtor's sign. This agent
was on duty. That agent was on duty.
We were together then looking for a home
that could grind our bones to make our bread.

Lionel

Feeling a kind
of jealousy for the heart
I redistribute
the basmati rice
in jars and close
the pantry door.

Outside a train crashes.
Inside a train crashes.
The christmas lights
you'd strung up, fall.

The conductors
take up residence
in our square patio
between two maples.
I challenge you,
one says, *I challenge
you*, says the other.

You challenge me
to say something,
say something.
Anything. Anything.

The Enjoyments

The apple trees are pink blossom today and all
streets are construction depots, hotrods tearing pavement
from yellow hell into the green pool hall heaven.
I slugged the cellulose capsule at the bottom of a Dixie cup
into my mouth, and now I can't stop complaining.
The foreman, conductor, engineer, Jerry From Across The Way,
and Phil The Accountant, sleep at their desks.
Their hands latched on suspenders like—*yup!*
Full as swans squatting in the paint factory:
an orgy of reds and greens and purples and greens
and reds and purples; if I only knew where to drive—
off a cliff, a roof, into caverns, then I'd drive.
My car is in the parking garage. I don't haul or cry for any victory,
or pray, oh I pray, but only in proxy: shit man,
I pray I left the car door locked. Things like that.

Mortar

The cars are red and white
and the mud sun slaps on them
like a sheath and sword
on a bumbling knight's waist.
The silence of morning globs
on me like a baseball bat through
a cob web. What am I feeling? How
am I feeling this? Wanting to find
the issue for why everyone I talk to
seems so fucking talkative,
the streetlights turn, the shops open.
In this hollow I call my skull
there is a shifting city like a Tetris tower,
each row taps out and sleeps an exhaust—
I tell myself the truth and know it
to be permanent.

The Purpose of the Thimble Inside the Cabinet

You are sitting in the laced light of a lamp,
that is how the quilt is made.
I am throwing a fit.
You are humming through a telescope.
I am throwing a fit in the gully
where potato salad fills the picnic
baskets of grandparents like computer memory:
gold and gridded with glowing theaters. A pair
of suspenders wrinkling in an attic, they hold
the greyhound's leash and walk. I fill
a cup up with well water and you take it.
I'm not fit for caring, the tractor says
in the field outside the kitchen, *No*, he says,
Not caring—carrying. Flat tires. Bent axles.
But the rain feels nice on my hood. I throw a bucket
into the sink and the night on my head.

A Foot and a Half

The cable television soaks into the couch,
infomercials for blenders and political panels
give way to pink trout bait floating in a stagnant pond.
How fast will the snow go down and hide in the ground
like A&D moisturizer on fresh ink? If all my past
is a mistake the season lays on me
as a stirrup shoveled in sleet outside a burning saloon,
I'll call it perfection. Just beyond the trees there
are more trees, blue haze, flurries of helicopters
and dripped wax of comet light. This is nothing to tell you.
There is no one to tell to. Save for the tiny
patch of vegetables, the bird feeder, bare now—
an off season hotel under an avalanching slope—
and my hair that I, one assured morning, chopped
off, and buried, and never dug up.

The Cartographer Knows
the Cryptographer Hid the Island

Our ship rolls into a reef.
The castle just past the storm's
edge crumbles on top of a harp
and all the knights gather,
at once, out of barracks,
tapping their helmets. Why do I
rub the carbon flecks in the kiln
between my fingers? Why is joy
a clay lock box in a melting snow field?
Work work work is a patience, but I trip,
I scatter, like a magician yanking a table cloth,
the dining set dishes crashing into the wall.
You need a little warmth. I need to know
my moat is real. We want a lot of air:
fill a quilted balloon, a woven basket to float
over a regiment. Also, carry a bright
copper ladle to serve up all the stew.

All is Bright

As for me, I say the sun
and the gong in my mother's
galley kitchen ring flat.
I talk to my life like the light
debates the dark—I can't see
the chimney flue. I can see
the chimney flue. At the foot
of my life there was a bell
ringing over an ant hill.

If you sit in the saddle
a horse will carry your
change. Hard tack in your
back pockets. The hare
twaddling by the brook.
A few deciduous larches
smack each other. At the table
in the tavern an empty glass
falls over, and I gasp, shrug.

Foul

I'm not as drafted on snow
as you are in water
and when we look at the barn
a goat falls out of the door.
Somehow this always happens:
a goat falls out of the door.
He must have been thinking,
he must be thinking, he must
have thought, there was a tall
rock, there is a tall rock,
there will be a tall rock,
he can, could, would, jump on.
As you said, this always happens.
Or was I saying it? We lean over
scratching our head together.

Anniversary

Today the floor shellacs my eyes. An oak splinter
is offered, free of charge, by the old floorboard
to a toe, slips in. How I'm lame as a shag carpet
under the mind's beige sofa in a drifted snow bank.
Today feels new with rain and under it I'm cold even
curling by a hearth flame like a kitchen bowl.
It is today: a clot in the flow of a life. The rain
and, as winter, cold. With splinters, in winter, lame,
blinkered, a soul rips and clots the rain. The bowl
holds bits of cereal, still-sogged. After you've vanished,
the boxes stay taped in the rusting storage units.

Fearing Another Break

The evening walks with a cane, mat white,
like a bombed out row of grain silos.
You lift your shirt and I lift my shirt.
These are the simple things we sell ourselves
in the catalog of our wanting.
The night slips on me like a cold sock.
Those feet, the stars say, I forgot we had them.
Yes, yes, they keep us in place
until we each say goodnight, goodnight.
A sky cut with SR-71s. A bed in the meadow
or is it a plot when between the power lines.

Proper Response

I lie. I really want to be cruel.
To yell into your filthy legs
the way a B-52 cuts clouds, fillets them
with warheads. As the grass frosts outside
the window your body is embalmed by seventy hands.
Take them all and make yourself up a ferry
that hauls from some primitive island Downeast
where the only office is a morgue, the locals
sipping brandy and ripping the ply for their plot.
Take them all and put them in your wicker basket
lined with honey and molded sourdough
that stales as the forest grows deep
and you skip along to a careless whistle.
I know there is a village inside the mind
populated entirely by janitors and when
everything is cleaned they must throw
the couches and desks out into the intersections.
They must watch the cars pile up from their bathroom windows.
All pain is a mouth that chews and spits.
The cruelty of loving is not forgiving.

What is Passing What Has to Breathe

Is a worm to answer
a squiggled question mark
in a censored sentence?
Sometimes all the cargo
your lungs ship up
to your tongue is raw,
airless, exported spices.
You say a few things.
Dig in, dig out of holes.
Then you die. Then *you* die.
And I do. And *he* does.
The cactus gets prickly around
the water spouts. Then you
are born in a room. Then they
are born in rooms. And I
am born in a room. And *he*
and *she* are born in a room.
People wave. People say yes.
People say no. Two cars stop,
and you cross the street. You walk.
People get taller. People get
smaller. Lots of hair. Lots of water.

Away From/It Floats

Along the cold window,
frost mirror, a mass
produced glass, I watch
the tomatoes yellow.
When the mail comes I
walk down. When the rain falls I
put on my slick. Grass
meadows out there blowing
behind the propane tanks
like rubber bazookas. Goodness
now comes to me in the broom
closet of the holistic institute's
music hall—head and hands—
playing my ribs like comet
trails through the harp I carry
for eons over our pretty years. Take it,
sharpen the arrow and build
a tree house of turquoise salt—please—
thaw the frozen branches, veins of Ötzi.

Vice

after Vasko Popa

in the house there is a kettle
and a moon over the attic
the wind spins the wind chime sour

there is milk in living and milk in death
is how the spoon shakes in the mug
with two arms and two legs

and a heart and many valves
pouring into dusted rooms with many lamps
milk in the living and milk in the death

I split a walnut on my scalp
now that's something to talk about
the shell just as in its place as the nut

Wilted or Not It's the Same Herb

The drying basil as good
as the thyme. The hearth as
built as the body.

The clouds made me and I
went out on a train
through the caverns

climbing stalagmites.
My blue eyes
they see themselves as I

see myself—fog, mist, a horseshoe—
as a Freman sees himself
in the water of the worm

making a hole in this
universe: a planet is
an infested apple. People

don't have the time
to be friendly.

"Two Minds."/ "But You Must be One."
for Claire & Ryan

I.

A pastor appears behind the mill
and the green pond half freezes.
The little hydrants in streetlight,
whirling quietly under their lids.
With white vapors a plane is a brief
ornament tinseled on the peak of a pine.

II.

I suddenly feel this happiness returning home
to friends painting by the fire in rare tact.
So, I can always be destroyed. If not by the lamps
going out then by the burn of such wicks.
I can always be destroyed. Opening my heart
I place inside it two small thimbles—a mouse in one
and in the other a monopoly column of salt.

What is It You Hear?

The faint frost over the sage
plants and each wing of the dragon
fly whistling as the sun heats.

Now that the fall is a cold folder,
coyotes leaping against the pavement,
the least we can say is, *man,*

I need some organic vegetables.
A mono switch on a Marantz amplifier
clicks. A record fizzes. Serrated silver

on a flat clove, white, the knife, flat
on the wood—this garlic peels
back like a pin cushion puffs.

I Always Think I Hold what I Love

They tell you your
circus is no longer
afraid, that the kettle drum
of ponds, whole
terrariums of capped lives,
are suddenly serene. They say
the dirt is only in
the cage, but there's
a body there. A cave
is only a lake when
the mountain falls
into water, hawks lava
while the Christmas
songs flicker through
the snow topped cabins.
Driving a brain southbound,
the moon separates
one shock of hay from
another stacked barrow.

In Sight of the Furnace

One succulent orbing on the window
and, likewise, the melt being itself outside…
melting. Cars at the top of Mount Washington
in the lot bumpered with declarative statements
in agreement of an achievement that each has made.
I take myself out of this only to get to the lift:
a tossed white storm trooper action figure
lost in the backyard snow. In the position
of the atmosphere, the heavenly zooms
into the noggin of the main character
to further the narrative and sidestep
the climatic pain of the hero dying
mid-story in a blaze. We have to often reheat.
We have to often move the ladder
while the siding of the house is complete
and a boy inside continues to cry on top
of a Persian rug. We have to often take
a shovel to the compost barrel and crack
the shells into steaming dirt, each callus
a little boiler—"What are you doing
down here?" pries the janitor, "Smoking
a goddamn cigarette," I say,
"What does it look like I'm doing?"

"My kin"

Shuffled coupons in her pocket book
and the pewter ice pick charm
soft digging into her wrist veins.
Doves link together on the steel
wire outside the stain glass church
window of Washington crossing
the Delaware—"A bizarre thing,"
she would say, "to have
in a Catholic Church." A bizarre
thing I think and think on a wooden
4x4 in the middle of the state park:
Why shouldn't I want to find the vine
impulse at the center of the split oak?
Why shouldn't I feed the index card dish?
And I am the wolf between a pair of pliers.
I turn to experience and experience,
it turns away. O fabled mist over
my singular sheet of flesh, thou art
dreams of many lives behind
electrical boxes and staring out
from incubators—green light to me
and white light to the rest, the watchers,
the keepers of joy and sorrow like gas
station attendants waving good-bye,
good-bye, to the band on tour
having just filled up and bought
nine Snicker bars and six Gatorades
and a rack of Miller camo cans.
After the mart closes up at eight

and Phyllis counts the unsold newspapers
wraps the day-old donuts for tomorrow's
discount bin—after the wood has been
stacked for next winter, the tarp thrown over,
it is so boring really to be alone.

Though I Long to Be Noah

A maple blows over the couple
carrying a desk out to the curb—cardboard
FREE flapping like a squirrel tail.
 I say *no* now to you in the garden
to myself. It's not that we die alone
which rubs me—that's ham and peas—
it's that, living, memories are—
populated pick-up truck bench seats
and broken hand clasps near stone walls,
a herd of sheep rolled about like a nimbus
in the meadow next to the highway tollbooth,
tomato stands rayed in the fall griddle sunrise.
I see you licking the orange in your mother's dress.
And to you your tongue hangs over each
corpuscular room—but, again, that's just me.
The premise is true though: many hands
touch a curb. In this market on third, someone
perhaps died lifting a bottle of Tide. Or by
the neighborhood creek, carrying a sinkable raft.

This Kills Me

I left my moped in the Emperor's
garden, orange ribbed over the bluish
patio. *No weeds!—No Weeds!*
the Emperor yelled, his window
sliding up, smattering down.
The palms lean against the stucco
like the ladders of the crushes in all
the Friday night sitcoms whining,
No weeds! at the slight odor of cartoons
on the flower bed. I left my jars of snow
on the driveway and dropped my shoes
on the Porsche. The freshly paved cul de sac
socked my feet coalish. *Imprecision all around!*
the Waste Management driver announces now
through his bullhorn. This kills me:
whether to stay in on my day off or go
out to someplace and stay in there?
On the stereo, just now, the band leader
tells me I can battering ram my life.
The Emperor feels his mahogany desk
shake under the shadow of his gold feathers.

Chlorophyll

Day always shorter than wheat
and the pleasure of instruction—
You do this. You do that—
carves the clay. Liquid
and a cup of victory
rises over the King's head
and his court falls down
like a poppy under moonlit
bunker fire. One way home
is to forget about it.
Another is to toss a six pack
ring set into a trash compactor.
This way the suddenly squashed
are with the walls like ferns.
A boat ganached with foam salt water.

Holiday Lawmaking

for Clark

Electric, the pounced tigers pop like static
along the play wheel, fully loaded on
jungle nip. I planned it: The ear to see
inside the shell. I briefed the grass
on its growth under the paw paw tree.
The stream crawled. The river stretched.
Nine ponds make a lake, I declared.
But everyone passed me as I rocked on
and on on my porch. Haven't the rulers
done enough? What rocky pasture is soft?
A thick ear sees the wind. My eyes
hear the glory of the coming of the slabs
in my yard. What I'm building are the byproducts
of a silver shoe—a shiny habit, a lopped buoy
by a strand of little lights not twinkling.
South of this house the fur keeps flowing
over boulders and people keep walking
in twos onto a manicured trail.

Many and Beautiful Things

You too! Birdish
to caw at a grape,
every day at this time
I go but I will not allow
to be wet with dew. My trust
is a Volkswagen Rabbit with rusted
shock towers loving some bug
eyes more than night maples.
I would present all reflectors on bicycles
as proof of the disappeared. Now,
lay my face on this cyclic honesty:
All this universe and not enough.

Strength

The train shakes with the spice rack.
It is winter and the thimbles have no clue.
A dove looks at another dove. A herd unmolds itself.
Were I to say something to you it would then be found.
All this findness and no reward for the loss.
All this thudding in the world's sap.
There can be one night and an angel trimming her skirt
for the Sun Ball. You too will love the music of shears,
forget about the ball, it'd be boring anyways.

Head in the Clouds

Even the finches
thrown up and tossed
around turbines like salted
boots in a mud room
cut his hair. The planes
chop at his earlobes.
A copter, on tour,
trims his mustache.
The body that lies
under water is at odds
with the density
of the gods: So does
his faith declare. This
giant. This fat-lipped
mountain, having been
wailed at by a NASA rocket,
you have nothing to look
forward to—moisture
and percolating swans.

The Hardened Garden

Too many pitch forks
not enough greens—
the earth has enough
fingers growing out
of the lapis dirt—know
what I mean, Roger?
Not that I asked for kale chips,
or corn stalks, or seeds.
I wanted roller skates.
I wanted the boxy tractors
to pull my shoes off
and guzzle the air.
The fog falls on me
like a leather recliner
on a cat toy batted below it.
My brother touches his side
with the one kidney left.

The Morals of the Cartoon

in the cardinals bush
running away with the nails
the cloud falls to the ground
without any knowledge
of the stars behind its back

Average Atlantis Map and Description

The gory sun
laughs on the bleached
moon while tide spills on
my jean shorts; tides
and tides river the earth.
Italy looks up.
The Congo looks up.
"This relationship
is looking to eclipse
the volcanoes and the cores,"
say the tectonic plates
yawning at the show. Under
the adobe sky, the blue huts
clouded with bakery smoke
are filled with fingers
turning to the passages
of esoteric cooking scrolls,
recipes for blood pudding.

The Hive

Turning trees into guns is easy; turning
mouths into walls or heads into pollen
not so in this morning rain.

Given a bouquet, the tree blushes;

patched bicycles fall over in rust
on the dirt roads. How convenient that
the alley, thinking, sprouts a hydrangea
like a dream in a brain, simple.

Dioramas of Bunkers

I.

spring into this drunkard talk
a volley of arrows break
on the ice river—block out
the sun. A cabin door locks.
Block out the wind. When I hear
that the postman will not be coming
I call the Sheriff. He calls the day
a wash. In a tin bucket outside
on the porch the water keeps
freezing and thawing a white towel.

II.

The sun is a lawyer.
Prudence is a cop car under
split wands in the car wash.
Adjustment was how we settled.
Virtue shoots a sewer cap
in the lattice clouded
dominion of your sky.
I slam the door in the tin
roofed house and move a shirt.

III.

Nagging in the lot.
Slums with rain puddles.
When will this earth
with its chains of sky,
scrapers, peelers, cottages,
emeralds of strip malls,

and latches of forests
wrap around my wrist
like the watch it is?

IV.
At the beach
a bear
sloshes in the surf
the tin eyes
of pedestrians
crossing the pale
winter ocean front
street. I, pain,
am dirt and flesh.
My body knots
up in the sunlight.
The undertow
rolls the bear.

V.
I cannot hide
from you
or your judgement
on me
like a bouquet
of nails polished
all of them
with an oak stem

VI.
You, in the other
house, what

are you doing?
Across the court
my hand veins
on your voice
through the ribbed
receiver of a tin can,
on your judgment
of how close the end
of the thread is from
your front door—neighbors
knock and bring sweets
and more hands into
the door frame.

Grounded for No Reason

after Alfred Starr Hamilton

I felt this was miniature
I felt this was impossible
I felt I sure can't do this
I felt I sure can't do this again
I felt I sure can't do this again and again
I felt of a small drop of rain that this rock is for the goats
I felt of a big turret on a castle that this arrow is for the teacher
I felt like throwing a little cloud over a small bed
I felt like tucking a pilot into a fighter jet

Poachers, He Says

after Robert Lax

Until	lion
exactly	spreads
noised,	fur
the	under
deer	each
keep	of
running	our
on	clocks.
the	*Time,*
green.	he
The	says,
sugar	*must*
grain	*adhere*
is	*to*
declared	*what*
dissolved.	*has*
Our	*been*
greatest	*decided*
most	*from*
noble,	*what's*
golden	*honored.*

Love the Woman in the Snow

But there are two of them
out at the lake
flipping cards over
and binding their
wrists to the table
in stubbornness and plight.
Inside, brussels sprouts
steam on the gas range.
I passed two horses today,
a galvanized bucket,
and tree cutters. Served
some food to customers
calling out their ticket
numbers on the white slips.
If I could undress
either of these two women
I would walk to my mirror
looking into my electric
irises and talk talk talk.

MDF

white tulips under the rifle
and the cherry fine
Virginia cigarettes
logo on the wall.
Square from the wall
you sand a block.
In the sky over the mill
the sun sands a cloud.
What new things we
learn when we knock
our heads on the glass,
sock each other on the lips.

Self Preservation is a Box Duct Wrapped with a Globe Inside It

sun that falls
the moon that falls
the cars that fall
over the off ramp
to the center
of a village
sitting on a water tower
the hawk is a game
and it is a losing one
to my eyes camped
under the ledge
of a setting forehead

Sweet Cyprus

The foil tinks along the road
past the bike rack bikeless
in ditch water. I'm not sitting
on a rock forever: a philosopher
holding a mushroom turns into
a birch tree or white ice over freezer
burned snow. When I came into
the pantry last night wondering
how dark the earth would be without
a human, my tears felt more
like sockets that the charging cars
take a day from. Then a month.
Then a year. The longer the night
the less time there is to watch a tree
on the edge of the corporate park
fall over onto an antique VW.
The less time there is to see
a package of Walkers shortbread
plaiding a pair of balled-up pupils
inside blue-orange irises.

Range

Folds in the clay
and the candle lit up
over the bath. I touch you
like the dirt over a root.
You touch me like the dirt
over a root. My face explodes
on your face. Your face
explodes on my face. They could
make a documentary about
just us and call it the reason
I think but do not say
into your mouth lipping
into mine. Outside the snow
explodes and three ducks
walk under a power line
and into the forest glow—going
there. Going there. Going there.
What is erotic in the sky, erotic
in the earth—to the clay slabs
wet you fold in a studio—the ground.
Our water. The dirt. Sky. What peak?

Love Poem

His towel is like his fed up sighs
when he says *Nah*
when her foot is wrapped with algae
and he's winding his watch
a truck or a train
her wet hair resembles
the light that makes two people
look like ripe pears dropped
on the hood of a lawn mower

Cut

> *after Sophie*

Why was this goat born? A missing leg, horn bump
caved in. The boy sings and flicks his hands on a roof;

new ships carrying spice capsize a league from the port,
and in opaque longing, we stir our striped coffee straws.

From the carriage house I throw a stone at a chicken
pecking an apple core—a handless planet of the proper gods.

The boy will never need to walk under a suspension bridge
where the dew rolls down the twined steel cables like fruit.

He'll find his healed rabbits and hydrated tea leaves
in hospital rooms air-lifted and dropped among forest dwellings.

It Figures

It's early. The snow
is (for the last time)
on the barn by the house
I move from (last
for me at least).
I'll wake up sometime
in the next three days
alone again. Sure. What
can the drooping
dog crossing the
cobbled stream have
to do with any of these
frozen tubers? What
passion I have is
a piston in an
alchemist's jalopy.

Ache

Having heard it, fragile through locals
and with intangibility
from foot in foot away from the crowds of fathers.

As close as Carthage now I ride to sell my lama. Hear that?
A woman is giving birth in a tub.
She is making the earth serious.
She is a blonde lily near a granite boulder.

Now at this parabolic hour
I listen for the orchestra of dinner triangles
playing Satie over beaver dams.

What in me do I seek? A tuner.

Nightcap

It is easier to sing this when the stairs
are more quiet. When the stars rise quiltish,
the creek of snow on the glass
not looking in the high-gloss transom molding,
as you walk over to your bed.
What you desire most is a spread:
registers ringing in empty hotel lobbies,
frameless Kincaid's, mossy traffic jam
in a highway of the eastern bloc,
and regular heartbeats per minute.
This world drips. The grass fiddles
between the rings of a paperboy
sitting with his Huffy on the neighbor's curb.
You could believe that—by the ceiling of the sky—
that you're ok, not cracked with a dozen
sunny side up eggs runny over a chow chows fur,
but then what would this cordless life refrigerate?

Alone and Happy

Everything keeps panging
over my skull. What place does
suffering hold in each meal I take?—
A mauve lamb in a creek filling up
after three months drought browns
like a patchouli dipped cotton ball.
The clydesdale lets out a sneeze.
The lamps rock in the glass wind.
"What a dweeb," a cook says
to a baker as they watch the lord
from the arched door drop
his chicken sandwich off the platter
onto his mastiff's slobbered head.

Rocking

for James Burton

According to the chair we are
sitting in, the floor is bowed and tipping,
trees peek over the window then hide.
The lawn outside, and the beach, bob
like buoys on the back of mud bogged pick-ups.
Saying an old story of my childhood,
I lost my place. Was this where I should have started?
At the mailbox? The chair tilts accordingly
and I stop accordingly. A bowl of cereal
sick with a heap of O's sogs in the counter
sunlight. The cat clock looks one way,
and then another, then another—tail
like the front lines along Flanders Field,
its eyes are pillboxes, open or not.

You are that and what is given

I wear my desert socks
I wear your watch with my face
I wear my suspenders with a paint brush and bucket
I wear your foot soldiers in my pocket looking
I wear my hills the first of the month
I wear your hammers and bang them on snares
I wear my self (I think you told that to me once)
I wear my watch with your face looking down

Cleanliness

A goblet of water is tossed
and the gas station lawn washes its blades.

Happiness lays down a cobble stone.
Contentment lays down a cobble stone.
The turtle lays down in its shell.

Nothing ends to look back.
Standing in line with a moon in my pocket,

I exchange it for a jack set of stars.
The woman holding pliers next to the lottery kiosk

pulls the name plaque off the office door.
The beginning drops over me.

Fire smolders on the hot plate
between church's chicken and my home:

The smoke is a cob web. The crawling is behind.
For whom are you weeping? My pockets ask my hands.

The bright light on the bay from the rig?
The sprays of rain over the heads

of lotus plants along the sheep pasture?
This earth which has to be so bright

that it hurts us, makes us shrivel,
and swallow, and breathe slower in its heat?

DIY

I'm reviewing your footsteps
in the sand. Well, this is a good place
to start. A wave came and you took it.
A barge floated by and I watched.
The sailor spit into a mug a bone
of a sardine too calcified to chew.
My sore shoulders gave out and I dropped
the umbrella. It is raining now.
Your head is bobbing one way down
the coast—in a jeep toward starlight.

Personal Statement

I'm going to bust the door down
and steal the motherboard. Going
to toss the virus into a stranger's
peacoat pockets on the blue line
into Crystal City. Then I want to make
something explode with delight
in the possibility of discovering a jungle,
or a building that's not yet been explored
because the architect was blind, wealthy,
and knew no one. What a practice you have,
I tell my disciples, spinning a canister of salt
on the checkered placemat. This is the hardest
of labors: leaves. This is the wisest of favors:
a pitchfork. I am convinced that this is the brightest
plan: an empty steppe with a herd of horses
dusting up in the rain like a leaking
drop ceiling over a homely spice rack.

I'm grateful for the following publications that have published versions of the poems in this collection:

PEN America Poetry Series: The Cartographer Knows the Cryptographer Hid the Island, Personal Statement, *Gramma*: This Kills Me., *Yes, Poetry*: The Hive, *Sundog Lit*: The purpose of the thimble inside the cabinet, *Hardly Donuts*: All is Bright, *The Puritan:* The Enjoyments, *Paper Bag*: Sweet Cyprus, The Hardened Garden, *Rochester Radar*: DIY, *Counterclock*: Head in the Clouds, *Nice Cage*: City Planning, Canal, Love the woman in the snow, Chlorophyll, *The Wanderer*: "My Kin", Cleanliness, MDF, A foot and a Half, "Two minds."/ "But you must be one."

The title of this collection takes its name from the Bob Dylan and the Band song, "Clothes Line Saga," off of *The Basement Tapes* (A Tree with Roots).

Gratitude as vast as the distance from Earth to Voyager 1 for Patrick Yagow, Freddy La Force, David Blair, Kai Carlson-Wee, Claire McHenry, Ryan Harrison, Elvis the Dog, Gracie the Cat, Sadie Brent, Wren Kitz, Christopher Messinger, Amy Sauber, David Rivard, Aaron Gerber, Keri Fernald, Gabriele Tise, John Greiner, Ben Lord, Erin Rooney, Alex Gorecki, the whole Chapel + Main family, Mom, Dad, Mark, Meagan, those whom I've loved and those whom have loved me.

Noah Burton lives in Burlington, Vermont where he builds coffins and makes coffee. This is his second book.

www.ingramcontent.com/pod-product-compliance
Lightning Source LLC
Chambersburg PA
CBHW030351100526
44592CB00010B/909